WOW
33 RPM
Flip to Play
Stereo
(Copyright Control)
MADE IN BRITAIN BY LYNTONE RECORDINGS LTD

MASTERMIND of
SOCCER - part 3
Flip to play 33⅓ rpm

Wobbly Sounds

Something

to

make

a

song

about!

Wobbly Sounds
A collection of British flexi discs

by Jonny Trunk

Four Corners Irregulars
№ 6

This is a

RANK AUDIO PLASTICS

SLIM DISC

WITH TWO PLAYING SIDES

A COMPANY
WITHIN THE RANK
ORGANISATION

Flexible Discs *Jonny Trunk*

For decades – certainly throughout the 1960s, 1970s and 1980s – the flexi disc held a unique place within the record industry.

It was thinner, cheaper, quicker to produce and literally more flexible in many ways and to more companies than a normal recorded shellac or vinyl disc.

We can trace the roots of the flexi disc way back to the first talking and singing postcards of 1903. Patents were claimed by 1904 for the 'enabling of persons, each provided with a gramophone, to converse one with the other by sending through the post a postcard or lettercard composed of paper or celluloid which has been previously impressed by the recording device of the sender's instrument'. These early postcards were made using clear celluloid with stamped-in grooves. And they sounded poor.

After the First World War, things started to change a little. A gluey resin called resorcinol-formaldehyde had been developed to protect aeroplane noses from heat, cold and weather. After the war, different uses were sought for this interesting new substance – including raincoat protection – and by 1929 music seemed to be a good fit. A British company called Worldecho started producing recorded discs for a postcard company called Tucks.

In the world of postcard collecting, Raphael Tuck and Sons were legendary. Started in 1866 in Union Street in London, Tucks produced extraordinary, influential and often quite experimental postcard designs, such as textured 'oilettes', which looked like oil paintings. In 1929 Tucks developed the idea of gluing the 3 inch musical disc made by Worldecho to the front of a postcard, with a whole pierced through both. These new 'Gramophone Record Picture Postcards' offered up new and bizarre clashes of traditional scenes or images and unrelated recordings, and proved to be quite popular. At about the same time, Durium, working with a similar gluey varnish (known as Durium Acetate) cemented to card, produced small, flexible weekly 'hit of the week' musical discs.

These were cheaper, easier and far quicker to make than contemporary shellac discs and proved even more popular. It wasn't long before Durium started moving into the commercial and advertising sectors with their new flexible discs.

By the 1950s, musical or 'singing' postcards were still very much in play, especially on the postcard racks at continental holiday resorts and tourist sites. Back home, even composers like Barry Gray (he of *Thunderbirds* music fame) were writing jingles and birthday ditties for the popular Melody Cards, which were destined for the post box. Companies such as Rank Entertainment started moving into the commercial recorded card and postcard market offering a variety of different recorded and paper options.

But by the early 1950s, one company in the UK was about to change everything. Lyntone. And over the next four decades, their presses would hum day and night producing countless charismatic little 'slim discs'.

If you are a record collector in the UK, Lyntone is hard to avoid. They are synonymous with flexi discs, and probably most famous for pressing The Beatles Fan Club series (see page 100). Over the years they were in business, they must have pressed millions of discs for thousands of clients. This is because nothing else could match the flexi disc or their factory for cheap, quick, tailored and often fashionable routes to promotion and the market. Exciting advertising novelties, wacky new dances, dull end of year reports, jazz for cheese, anything could and would go on a flexi. And Lyntone would make it fast, and make it well.

As for distribution, once made, your flexi could be given away in cereal packets, posted to you by Reader's Digest, handed out at exhibitions or shows, swapped for coupons at petrol stations, stuck anywhere in magazines or comics, sent off for with packet lids, but very rarely sold.

Many countries around the world have their very own 'Lyntone' and huge flexi disc history – especially the USA. But for this book, we have just focused on flexi discs manufactured in the UK. Most, as mentioned, were made by Lyntone, but a few were handled by Rank and other smaller plants. As you will see from this very first book on flexis, they were used to promote everything from pop to porn.

As any buyer of second hand vinyl will know, flexi discs have an interesting reputation. They can look fantastic. From their sleeves and descriptions they often promise excitement, sonic oddness and period perfection. And actually they are remarkably durable. But they nearly always sound terrible compared to their more solid but brittle vinyl counterparts. Over the years I've developed my own ways of playing these wafer thin recordings. First, you have to find a smooth surface as most record players can have a ridged platter and here even a soft slip matt will be of no help at all. So, I find a smooth piece of medium weight cardboard and cut it to roughly fit my revolving turntable. Then I cut a small hole in the middle and place it over the spindle. Next, I place the flexi disc over the spindle and rest it on the cardboard and instead of using the recommended 'small coin', I place a metal record centre – the type you can buy if you are an enthusiastic 45 collector – at the centre of the disc. This flattens the centre, stops the disc from slipping and the cardboard provides a smooth firm base causing no interference as your needle rides over the tiny flexi grooves. The sonic results are superb. Or as superb as you can get with a flexi.

It's because they are so often hard to play properly that many are avoided, left to wait in secondhand shops and car boot sales until a flexi enthusiast (like the ones whose collections make up this book) ignores the sonic shortcomings and just wants to briefly wallow in this simple, wobbly little wonder.

As for the sudden demise of the flexi, production ground to a halt as the CD invasion of the 1980s took hold, and anything to do with vinyl or certainly records of any shape or form were quickly forgotten. Now, decades on, we have a new vinyl revolution, and pressing plants around the world are once again trying to perfect flexi manufacturing. I've seen a few as freebies for music magazines in the last 12 months, but I somehow think the day of the flexi has been and gone.

Thanks for listening,
Jonny Trunk

Huge thanks to the flexible flexi collectors who helped put this book together by donating their own amazing collections to the cause: DJ Food, Andrew Beedell Coram, Shane Quentin, Jez Randell and Jonny Trunk. Special thanks also to Sandra Lynton.

Lyntone *Alan Dein*

Lyntone was founded by Paul and Lya Lynton at the tail end of the 1950s. Paul was born Paul Lichtenstern in Budapest in 1922. He grew up in a comfortable Jewish home in Vienna. After his father was arrested and transported to Dachau concentration camp, Paul managed to get to Palestine in 1939. He had a tough time at first, sleeping on benches, until getting a job as a musician in Tel Aviv. He played Double Bass and the Accordion.

He subsequently joined the British Army, and became a member of the ENSA swing band, where he entertained troops in North Africa, India and Sri Lanka. He was demobbed to London in 1946, where he worked as a musician and a theatrical agent in Soho. He specialized in cabaret acts. Just before Paul married Lya in 1950, he changed his surname to Lynton – though retained his old nickname 'Bobby'.

His work diversified into TV advertising, in the early days of Independent Television, where he had a spot promoting various brands. This shaped an interest in postcard style giveaway records. He obtained the licence to manufacture German-style postcard records under the Lyntone name, and the US Evatone Soundsheets, in the UK.

In the early 1960s, Lynton's Audio Plastics company was bought out by the Rank Organisation. The company was then rebranded and continued to manufacture what Paul Lynton described as the 'slim discs'. His pressing plant was originally based at 46 Penton Street near the Angel, Islington. In 1970 Lyntone expanded their operations, moving to much larger premises in Wedmore Street in Holloway.

During the 1960s and 1970s the Lyntons had amassed an ever-increasing workload as the nation's manufacturers eagerly promoted brands with their very own audio recording, on disc. There were slim profits in flexis, so it was a matter of turning around as many units as possible. Lyntone was a fully independent, family business, with Paul overseeing all the technical and engineering processes – and Lya looking after the sales department.

The Lyntone factory, as featured on a promotional flexi disc sleeve.

The Lynton's were also helped out by their son Robert and daughter Sandra. Several amusing stories still stand out in the non-stop era of record production. Sandra remembers that when Lyntone got the order to produce the Hare Krishna giveaway records, part of the deal was that her father spent a day in religious contemplation at the Krishna temple.

Sandra also recalls the time when Lyntone discs appeared on the hugely popular *That's Life* BBC TV programme – after batches of flexis ended up being sent out in the wrong covers and labels, so kids-themed flexis ended up with *Readers Digest*-sponsored music on them!

Also, between the 1960s and the 1980s, Lyntone pressed the famed *Private Eye* flexis that were given away free inside the satirical magazine. Paul Lynton made sure that *Eye* signed a contract that Lyntone would not get into any legal trouble if the oft-sued publication ever got taken to court about the content of their discs.

By the late 1970s after the punk explosion, Lyntone began pressing more of the traditional 7" & 12" vinyl, and bands would venture directly to the Wedmore Street plant to get their DIY records pressed.

After three decades of producing millions of 'slim discs' for thousands of clients, Lyntone eventually called it a day in 1986.

The Secret of Beauty and Healthy Sleep, 1932 (Durium Products, E 5001). 78 rpm Durium disc. Still trying to find out what the secret is, as this very small disc did not play due to the plastic cracking off.

Dansette, date unknown (Hardy-45).

The Washboard Birthday Special, 1950s (Melody Cards).

Christmas Greetings, 1958 (CS 603).

We'll Gather Lilacs, c. 1958 (Lyntone).

Here's Wishing you a Happy Birthday, 1958.

Pretty Talk, The record that teaches your budgie to talk, c. 1958 (Lyntone).
Two sided instructional disc – simply the same words and phrases ('clever little boy', 'give me a kiss' etc) repeated endlessly so your budgie may catch on, but actually makes you think the record is scratched.

Triumph Herald Presentation Music, Triumph Cars, 1959 (Lyntone).

You're Never Alone with a Strand, 1959 (H-W-33).
Two sided promo for infamous failed cigarette commercial – in 1959 no one wanted to be alone, so no one bought the cigarettes. One side is a turgid address by the company chairman, the other a fine, smokey and slightly twangy blues sung by Jimmy Lloyd.

Michelin 'X' Tyres are different, Michelin, 1960 (JU 59B).
Double-sided 33 rpm Caribbean-style ditty selling safety tyres:
a case of nice idea, nasty execution.

Lustre Crème Shampoo, c. 1960 (LYN 50).
Complex one-sided dance-step instructions for new shampoo,
of course they go together really well.

Disc-course on Beauty, Elizabeth Arden, 1960 (LYN 97).

A Personal Call from Squires of London, 1960 (LYN 114).

Wherever You Go... You Look Better in Arrow, 1960 (LYN 115).

Get into the Britax Safety Habit, 1960, Britax (LYN 115, repeated catalogue number).

Wise Birds Have Oil Fired Nests, FINA, 1960 (LYN 124/125).

Spring Cleaning Time, Sound Marketing Ltd., c. 1960.
One-sided trade disc for the sound marketing service highlighting new adverts to be played out in supermarkets. Brands include Super Squeezy, Ship Household Matches, Brillo Soap Pads etc.

Serenade, Cliff Richard, 1960 (Rainbow Records, HUMA 1).

The Mackesong Medley, music by Johnny Johnston
and The Keynotes for Mackeson Stout, c. 1960 (LYN 126).

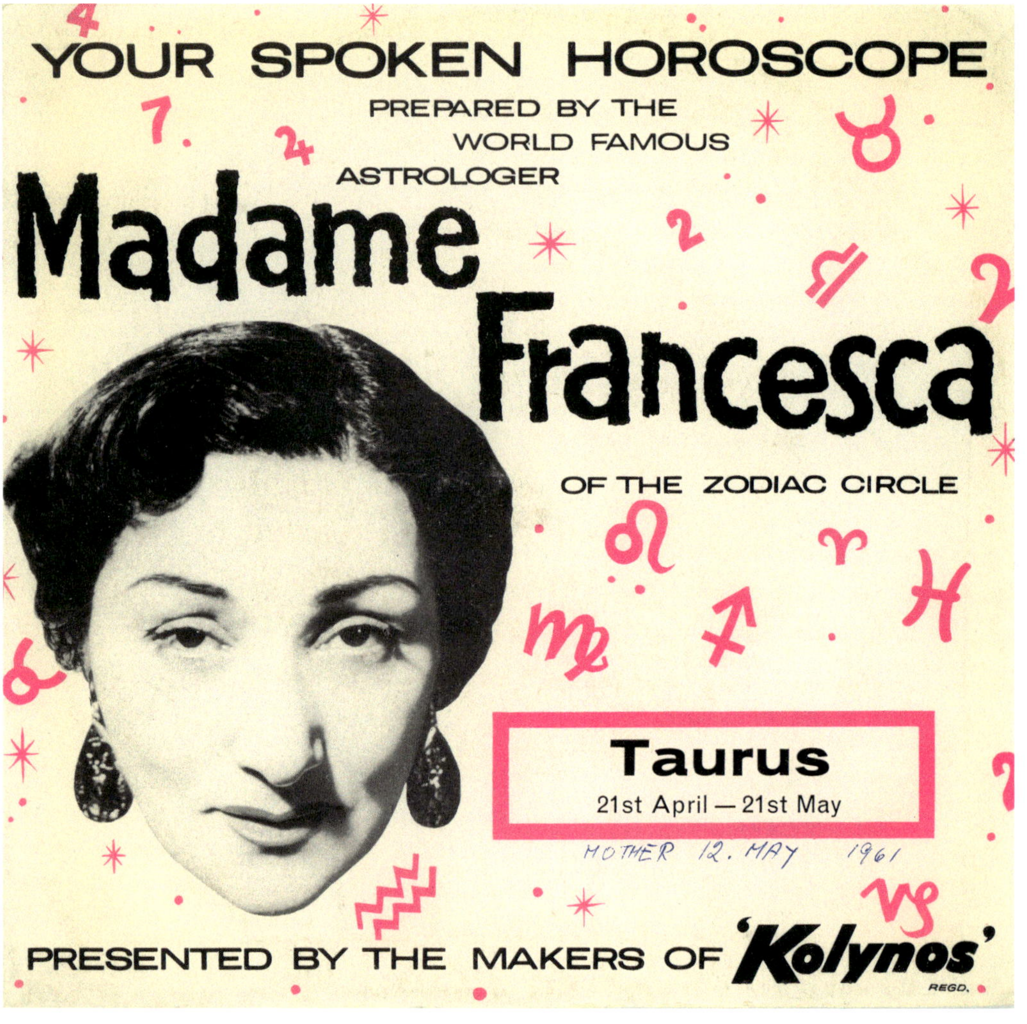

Madame Francesca, Kolynos, 1960.
Two-sided disc from Kolynos – a toothpaste manufacturer. One of a series of twelve discs, one for every sign narrated by the world famous astrologer – who was in fact Phil, the brother of crime writer Margery Allingham. All backed with a light Quickstep by Stanelli and his Quartet. Issued in 1960 for 1961.

Listen No Buttons, Myers, 1960 (LYN 139).

Chaseside Motors, 1960 (LYN 143).
One-sided 33 rpm card recording – four over-written theatrical mysteries guiding you towards buying a new Ford.

A Message from the Chairman of Bryant and May to Commemorate 100 years of Making Matches 1861-1961, Bryant and May, 1961 (LYN 147).
Incredibly dull freebie for the matchmaking staff. One-sided, thankfully.

Françoise Garrigues Talks Fashion, Peter Robinson Fashion Store, 1961 (LYN 152).

Excelsior Corsetry, c. 1961 (LYN 163).
One-sided 33 rpm card disc. Drab advert-based disc telling you of the 37 insertions of a new corset advert in the press. But really sounds ace at the wrong speed.

French Lesson, Odhams, 1961 (Printed Sound Ltd).

Tango Rock, Dave Carey & Barbara Moore for Tango, c. 1961 (LYN 195).
One-sided 'glued' 78 rpm – catchy jazz tango for orange fizz and swinging parties.
Sung by vocal legend Barbara Moore.

Diabinese. The control of Diabetes, A Pfizer Medical Recording, 1961 (Fon-176-1/Fon-177-1).

Family Planning with Nurse Drew, W.J. Rendell, c. 1961 (LYN 198).
Issued by W. J. Rendell Chemists. A one-sided 33 rpm flexi. Nurse Drew talks candidly about birth control and 'unwelcome interruptions' when using sheaths etc. Supplied in a brown paper bag.

HMV *Junior Joyride*, c. 1961 (LYN 199/200).
Steve Race takes us on a trip – promoting HMV Junior spoken word books on record.

Forge Ahead, a Personal Message to you from... Tractor Group, c. 1962 (LYN 216).
One-sided talk about tractor sales. Unfathomable, unless you had a tractor in 1962.

Supercar Club, music by Barry Gray, 1962 (LYN 249/250).
National Garages tie-in with the classic early 1960s Gerry Anderson *Supercar* show. Great short story with Supercar landing at a National petrol station. Full cast, SFX and Barry Gray music. Features the lively 'Supercar Twist' sung by Mike Sammes on the B Side.

Milk! Tracks from a selection of Milk Television films, W. S. Crawford, 1962 (LYN 313).
Series of catchy, twangy and beat driven scatty numbers all about drinking milk a lot.

Cave Girl, Clarks Wessex 'Top Twenties' Shoes, theme music from the film, composed by Bill Shepherd and performed by Bruce Turner and his Jump Band, c. 1962 (LYN 208/209). Two-sided promo for Clarks shoes, as featured in their two minute 'Cave Girl' advert set in Chislehurst Caves. Bruce Turner supplies the lively trad jazz for the ravers.

Slimming, Woman and Beauty in association with Leas Cliff Products, c. 1962 (LYN 222/223).

The Dutacq's Camembert Melody, Crowsons 'The fancy cheese people', music by Eddie Gurney, Grisham Farfel and his Music, The Rita Williams Singers, c. 1961 (LYN 225/226). Two-sided catchy calypso style number all about cheese, and remarkably, not that cheesy. Two versions: one by Grisha Farfel, the other by the Rita Williams Singers.

Prestige Sky-line, music by Chris Barber, 1961 (LYN 235).
Lively one-sided trad jazz for modern kitchen equipment. From the Prestige TV commercial.

Mayday, Mayday, Mayday! (Rescue by Radio), Pye, 1961 (LYN 238).
'Here is a story of safety at sea. This record tells, very briefly, how the owner of the ten-ton auxiliary Bermudan sloop "Caragh", his wife and their dog were nearly wrecked in a force 10 gale off the West coast of Scotland.' Smooth voice selling the Pye Marine telephone, useful when the going is rough. Includes some poor reenactments of disasters at sea and some real ones too.

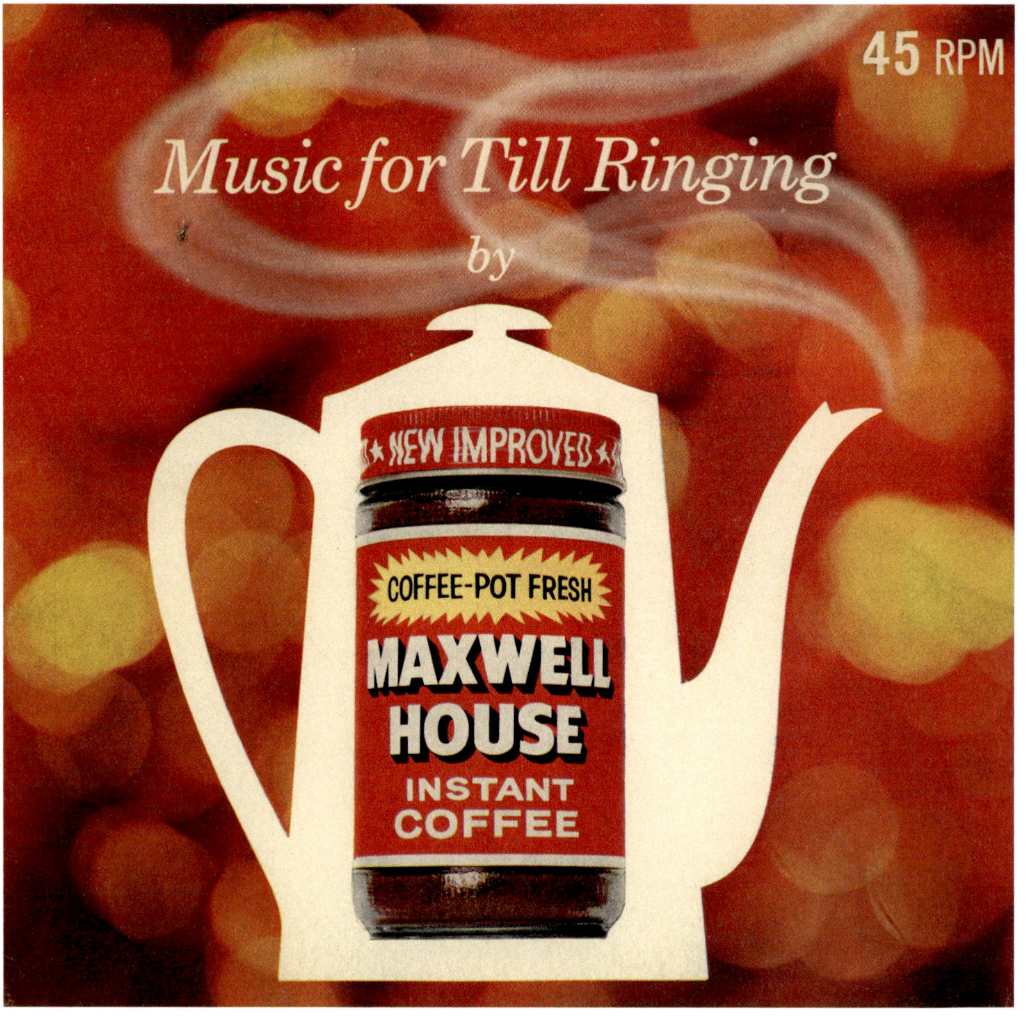

Maxwell House, 1962 (LYN 246).
Charismatic Caribbean tinged percussive number, followed by a retrospective of gay and lively Maxwell House jingles and product development.

Eat and be slim with new Limmits, Leas Cliff Products Ltd., 1962 (LYN 253).

Lost, Phyl & Grisha Farfel for The National Society of Mentally Handicapped Children, 1962 (LYN 315).
Unusually presented one-sided disc, glued to a fold-out sleeve, that folds back on itself to create a smooth surface to play the disc on. Brilliant really, should have caught on but didn't. Features a well-sung torchy blues by Phyl Farfel.

PLEASE REMEMBER
WHEN THERE'S SOME - ONE WHO CARES, WHAT BE - COMES OF YOU, SOME - ONE TO WHOM YOU CAN
RUN, SOME - ONE WHO'LL GUIDE WITH A FRIEND-LY HAND, WHEN-EV-ER CLOUDS ARE HID-ING THE
SUN. THESE ARE SUCH LIT-TLE THINGS, BUT WE'RE ALL CHILD-REN AT HEART,
AND IT'S FROM THESE LIT-TLE THINGS, THAT WE FIND THE COUR-AGE TO MAKE A NEW START.
SO WITH A CHILD WHO NEEDS EX-TRA CARE, LET NOT THE NEED BE DE - NIED,
IF THERE'S A WAY IN WHICH YOU CAN SHARE, TO LET THEM KNOW YOU'RE THERE, RIGHT BY THEIR
SIDE, DON'T WASTE A MO - MENT, DON'T COUNT THE COST, FOR WITH -
OUT YOUR HELP THEY MAY BE LOST, LOST WITH - OUT YOUR HELP.
(RIT--)
Printed by Bentley & Co. (Printers) Ltd., Hitchin, Herts.

Sweet Dreams for Swinging Sleepers, Pleetway Pyjamas, c. 1963 (LYN 344).
One-sided trade disc, promoting the up and coming Pleetway Pyjamas TV commercial.
Enough to make you turn it off and go to bed.

PLEETWAY
PLEETWAY
PLEETWAY
PLEETWAY
PYJAMAS PYJAMAS PYJAMAS PYJAMAS
zzzzzzzz

Marching Union, Acker Bilk for the National Union of Boot and Shoe Operatives, 1960s.

Tunes from TV *Times, Wiggle and Giggle*, Cliff Adams and his singers, 1963 (LYN 368/369).
Two-sided promotional disc celebrating both Cliff Adams Cha-Cha Chas
and a new series of TV *Times* 'insane' commercials, starring Clive Dunn,
Fenella Fielding etc.

Chef's Discs: India, record recipe for chicken curry and rice, presented by John Bluthal, 1963 (LYN 493/494).

Meet 'Mr Millington!', Sydney S. Bird and Sons, 1964 (LYN 548).
Ron Millington plays the groovy Bird Organ – a keyboard with strange, spooky ethereal sounds.
Sadly an annoying man talks rubbish poetry about the organ over the top.

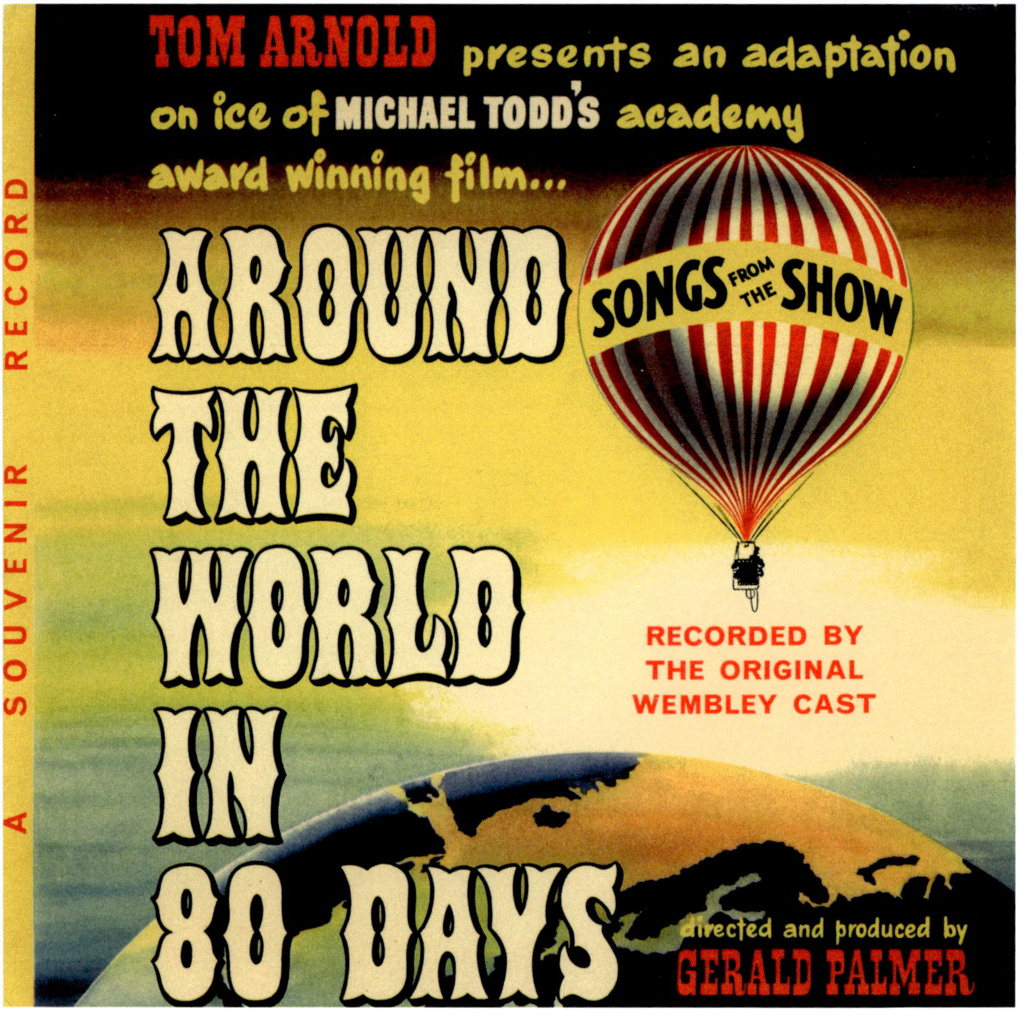

Around the World in 80 Days, Gerald Bourne Production, 1963 (LYN 506/507).

Recorded for George Newnes Ltd., c. 1964 (LYN 547).
Cunning argument and interviews made to talk you into buying a set of Chambers Encyclopedia. One-sided.

That Wonderful Worthington Taste, music by Johnny Johnston, early 1960s (Hardy Records).
One-sided trad jazz-based advertising song by Johnny Johnston,
a legend in British advertising jingles.

Play Time on the Hammond Organ, music by Keith Beckingham, 1964 (LYN 564/565).

Here Comes Courage, 1964 (LYN 570/571).
German 'bierkeller' style number for this classic of British beers, with a sales voice-over in the middle. On the B side we have the 'Courage Twist', a brave attempt at a pop dance number, restricted by the honky-tonk piano that was obviously a demand from the client at the time.

The Hand of Mary Constable, Paul Gallico for The Book Society, 1964 (LYN 572).

Bike Beat, Raleigh, 1964 (LYN 573/574).
Two-sided. Groovy go-go sounds by the Rally Rounders, AKA The Outlaws, a band that featured Ritchie Blackmore. Tunes produced by the legendary Joe Meek.

Songs for Swinging Voters, Conservative & Unionist Central Office, 1964 (LYN 586/587). Various well-executed music-hall style numbers poking fun at Labour, from the Conservative and Unionist Central Office, for the 1964 election. Running over two sides, it's not as funny as it wants to be, but then again the material is quite limiting. In fact, the title is the best bit.

Get Hep with the Hepworths, Johnny Johnston and The Keynotes for Hepworth's Handcut Tailoring, date unknown (Printed Sound Ltd).
Two-sider with twangy pop number on the A Side, a personal message from CEO Norman Shuttleworth, about hand-cut tailoring on the B.

Take a Spin, commentary by Barry Alldis, date unknown, Sound for Industry (SFDK MI J).
One-sided, one-dimensional jazz-tinged promo for 'Sound-Kard' flexi discs.

Young Londoner Shoes, date unknown.
Fine single sided 45, jazz infused swinging shoe advert to promote new Spring promotion and cinema advert. Apparently 'you can step out of the blues in Young Londoner shoes.' Of course you can.

Curry's Christmas Party Record, date unknown (AR-CI).

Bulmers' Do it Yourself Comedians Kit, Bob Monkhouse for Bulmers, 1963 (SHOL 266). Surprisingly clever 'party' record produced by cider maker H. P. Bulmer. Presented in a gatefold sleeve with a leaflet of instructions, this disc allows you to enter into a double act with Bob Monkhouse – you simply talk in the gaps he leaves. Some entertaining use of the medium too, by putting a deliberate 'scratch' into the record that Bob thanks you for bumping on.

Enjoy Dancing, Talking Book Company, date unknown (1252/U).

Dialect Records: American Speech – Middle-Western/Southern,
The British Drama League, 1964 (LYN 684/685).
This is one of 16 records in the dialect series, with 12 British and 4 American issues.

Face to Face, Mills & Rockleys Screen Printing, early 1960s (CHA 4801).
Strange one-sided 45 rpm promo disc for screenprinting based on making no mistakes. Fairly ineffective examples of 'classic' mistakes and bloopers from speeches used to press home an irrelevant point.

It's Carpet Showtime, 1965 (LYN 702).

Brillo Clearway, 1965 (LYN 762).
One-sided. Spoof talk by Professor Pulldeplugout talking about plughole noises.

A Recorded Talk by Nurse Drew on Family Planning, W. J. Rendell, c. 1965 (LYN 773).

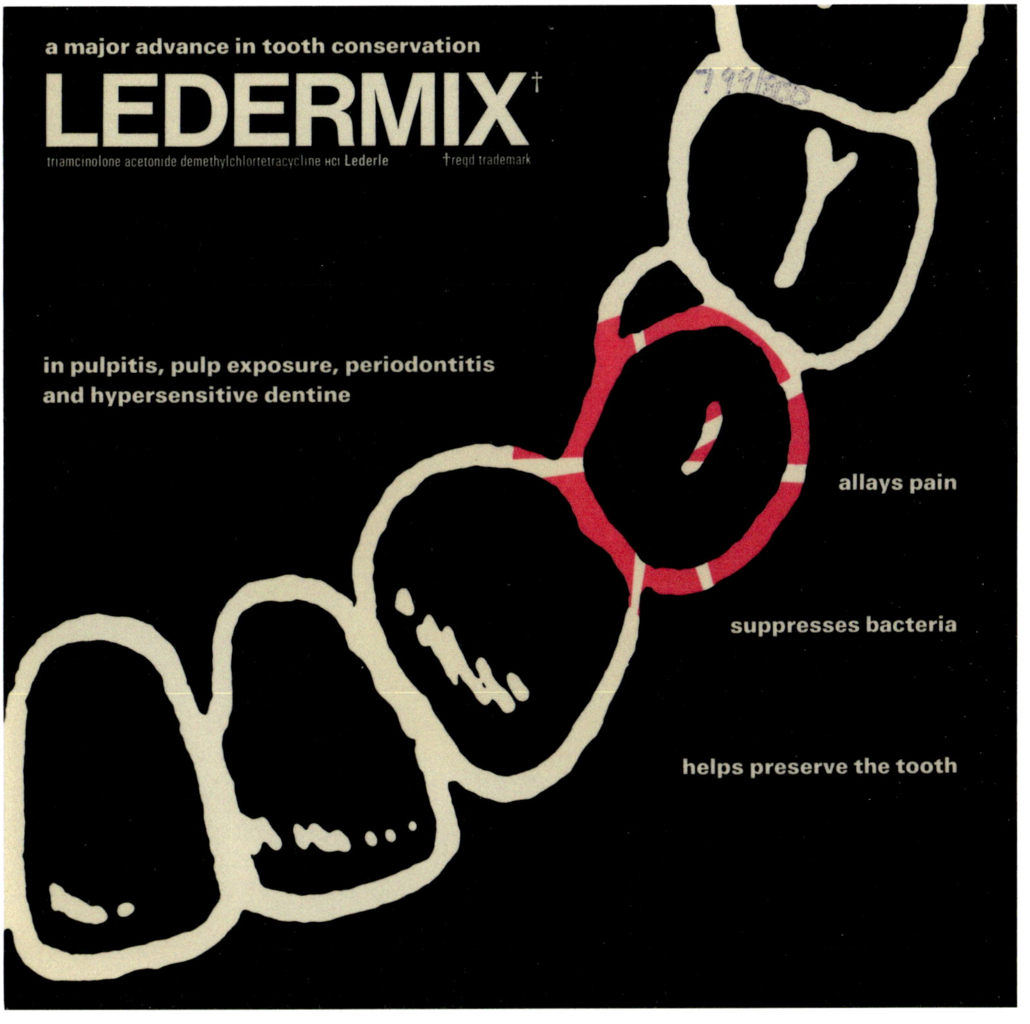

Ledermix, 1965 (LYN 799/800).
Two-sided. Serious bit of scientific promotion for new dental product from Lederle Labs.
Includes some slightly dubious interviews with dentists.

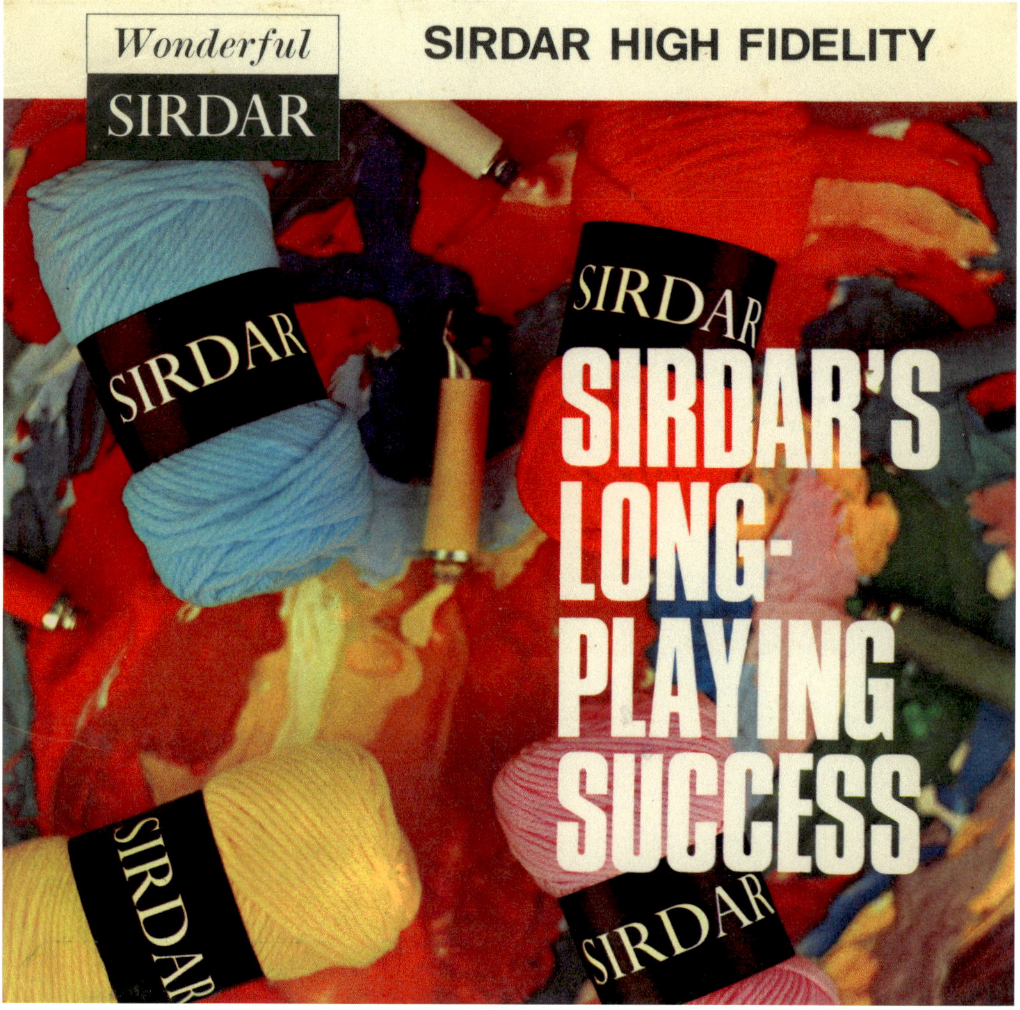

Sirdar's Long Playing Success, Sirdar, (LYN 830).
Quite beautiful one-sided sleeved card disc from wool manufacturer. Set of intriguing facts about i) grammar and then ii) the company, all punctuated by vibraphone.

IRDAR

Clearasil's Clearway to Skin Health, Clearasil, 1960s.

Harry Edwards Speaks from the Spiritual Healing Sanctuary, 1965 (LYN 927/928).
This runs over two sides at 33 rpm, so it's one of the longer flexi recordings out there. Sadly it's just Harry Edwards talking – he may have the gift of healing via spiritual attunement, but he was back of the queue when it came to narration.

The Salvation Army International Centenary Celebrations, London, 1965 (SHOL 529). Two-sided Salvation Army souvenir, with a 'handful of delights' from the musical celebrations – brass bands and folk.

Soundtrack to Shell TV *Commercials*, 1965 (LYN 772).

Teddy Bears' Picnic, Sugar Puffs, 1965 (LYN 929/930).

Clarks Shoe Beat, c. 1966 (LYN 997).
One-sided glammy, slammy number with a simple rousing 'Clarks' as a chorus.
Some good, heavy guitar to go with those heavy wedges.

Here's an invitation from Wavy Line for a fabulous weekend!, Audio Plastics Ltd., 1966 (LYN 980).
Disc inviting you to the up coming Wavy Line 1966 grocery sales conference.
It's at Butlins, Clacton On-Sea, 16–18 April. See you there.

Durabolin, 1965 (LYN 959).
One-sided 45 rpm chat about uraemia and the new drug therapy.
Several big serious words here you will only understand if you are in the business.

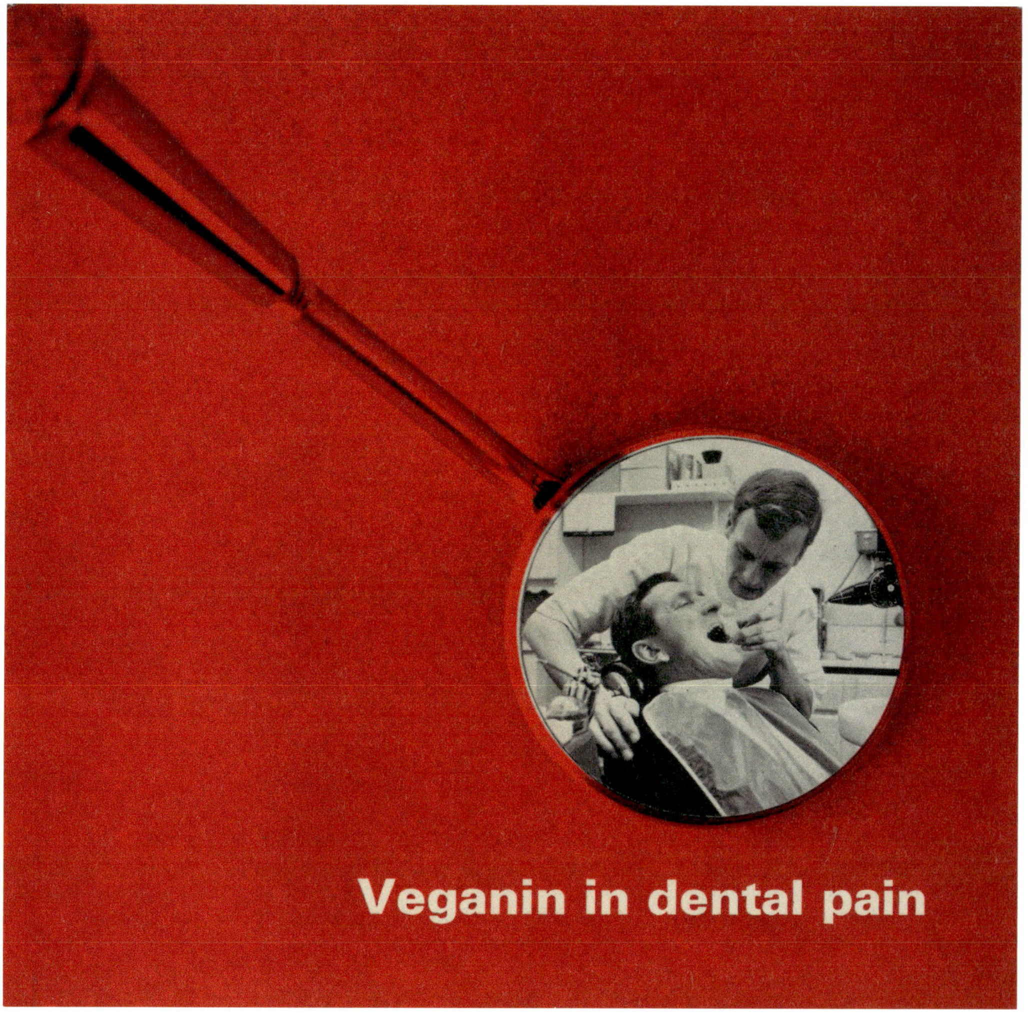

Veganin in Dental Pain, 1965 (LYN 961).
One-sided disc about painkilling in dental care. Yes, you've guessed it, Veganin provides that post-dental pain relief your patients want.

Neutrogena Soap, 1965 (LYN 971).
Early promo for new Neutrogena soap. Four new American-style commercials giving away free samples on Radio London. Must have worked as it's all over the place now.

AP *on* LP, Allardyce Palmer, 1967 (LYN 1000).
One-sided trade disc by AP Industrial; a painfully boring lecture promoting their exciting advertising agency credentials.

When it Comes to the Crunch (It's Smiths IT IS!*)*, Barnes/Adams, 1966 (LYN 1021/1022). Groovy two-sided 45 rpm – mod meets Smiths Crisps. Instrumental and vocal sides. Certainly one of the better crisp-based songs out there.

The Dippity-do, Toni/Mecca, 1967 (LYN 1092).
Late sixties hand-clapping poptastic number based on a new hair gel.
Free from Mecca Promotions. One-sided.

She's Got It!, Cortina, 1967 (LYN 1104).

The 'Ayes' Have It, Richard Murdoch & Kenneth Horne, Presented by The Institute of Opthamology, 1966 (LYN 1119).

Practical Electronics, October 1967 (LYN 1243).
Legendary one-sided 33 rpm disc free with *Practical Electronics* October 1967. Introduced by electronic pioneer F C (Frederick) Judd. Basic sounds from electronic sources – pure tones, white noise, ring modulation, filtering, recording techniques and a ditty. Total joy.

Velb, 1969 (LYN 1793).

The Care of Your Real Hairpiece, Steiner, 1967 (LYN 1245).
Full instructions over one side on how to care for your hairpiece, presented in the dullest way possible by the female narrator.

Party Noises, Cadbury's Whole Nut, 1967 (LYN 1358).
One-sided, die-cut sleeve. Cadbury's Whole Nut Party Political Broadcast.
Not funny then, not funny now.

Party noises
45 RPM
LYN 1358
Issued by the Cadbury's Whole Nut Party
All rights of the manufacturer and the owner of the recorded work reserved. Unauthorised public performance broadcasting and copying of this record prohibited. Made in England by Lyntone Recordings Ltd.
Not for sale.

Three Blind Mice, Roger Hyslop, Pop Outs, c. 1968 (LYN 1391).

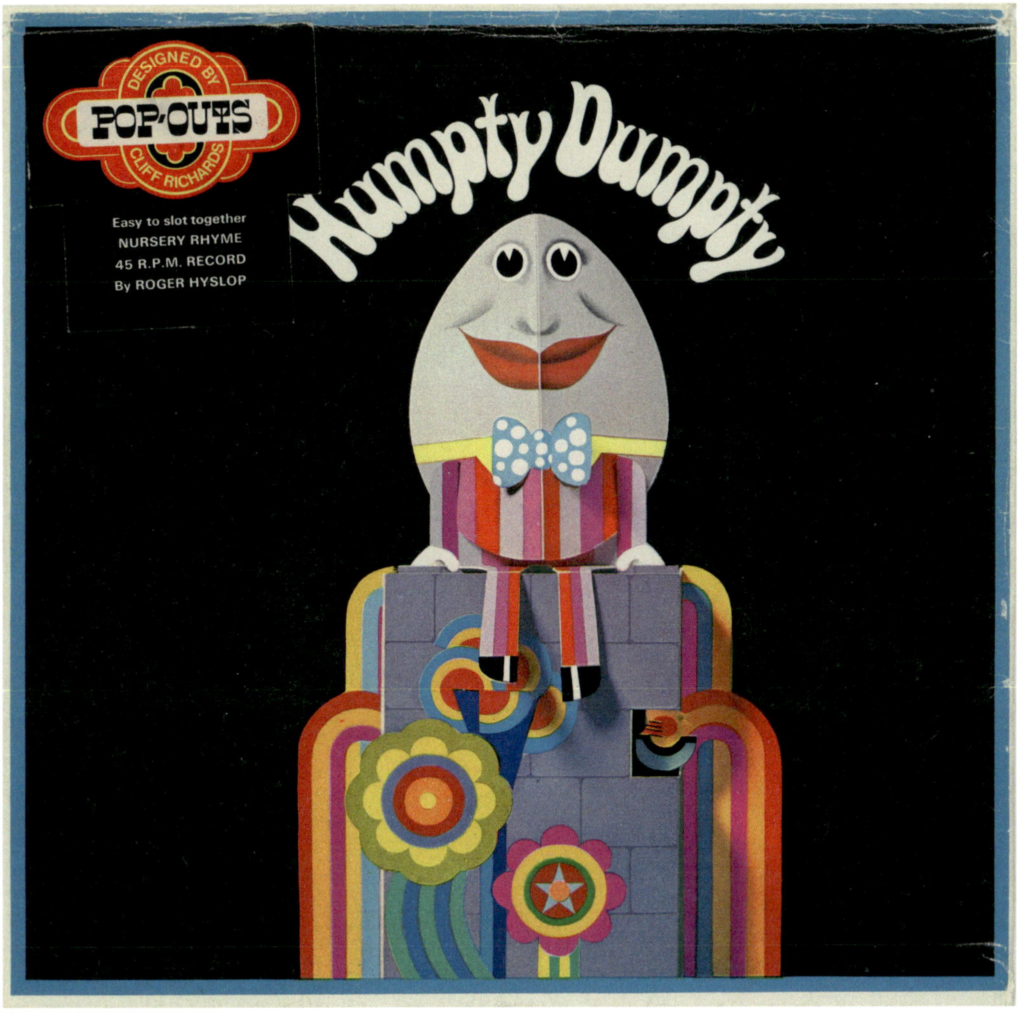

Humpty Dumpty, Roger Hyslop, Pop Outs, c. 1968 (LYN 1392).

Young London Permissive Paradise, 1969 (SFDK 31/32).

A very rare two-sided flexi issued by Harrap, the book publishers. We believe this was pressed just as a giveaway, probably for bookstores and record shops, to promote the new (and now classic) book of the same name by Frank Habicht. Habicht, a German stills photographer who'd been working with Roman Polanski and Brian Forbes in the UK, spent a year observing London youth at rest and play. He magically captured the sights and vibes of what was to become one of the most extraordinary, influential and quite mad years in global pop and fashion culture. The book was issued just once in 1969 and has since become a keystone of photo journalism and style. Side One of the disc has documentary style cut-ups of Carnaby style street interviews, narrated by Norman Tozer and including reviews of the book's photos by Radio 1/Caroline DJ Emperor Rosko and the then rising music executive, Jonathan King. Side Two features a specially written song – 'Permissive Paradise', performed in a pop/psych style by 'The Pleasure Garden', AKA The Iveys, AKA Badfinger.

YOUNG LONDON
YOUNG LONDON
PERMISSIVE PARADISE
EMPEROR ROSKO
JONATHAN KING
and
THE PLEASURE GARDEN

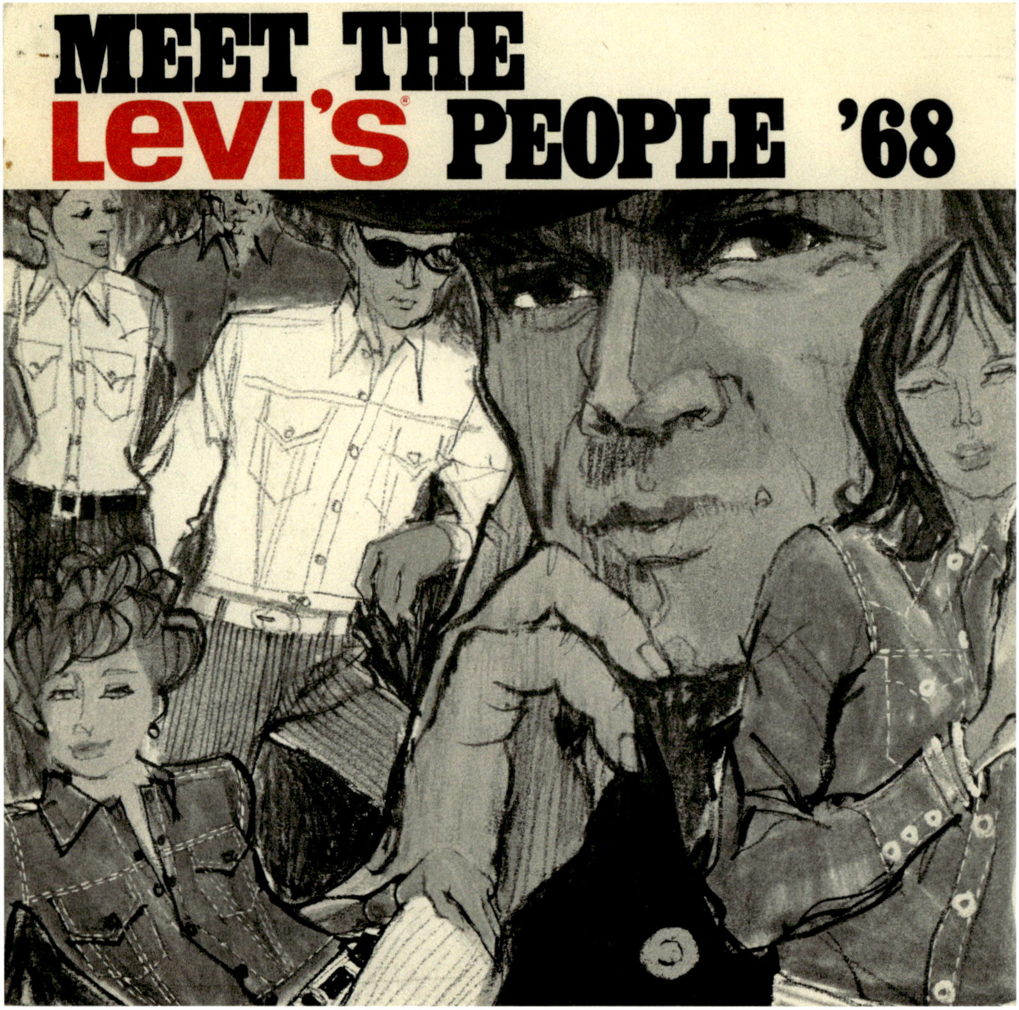

Ask for Levi's by Name, Chappie D'Amato, introduction by Benny Lee, Levi's, 1968 (LYN 1396)

Mood Elégance, Avon, 1968 (LYN 1409).
A new perfume by Avon promoted by this new one-sided 45 in a fine sleeve. Great orchestral score and superb period voice-over by God-of-1960s-voice-overs, Patrick Allen. Came with a trial sachet that in this case was missing so we could not experience the subtle 'whiff of France' and 'captured mood of elegance'.

The Loneliness of the Long Playing Record, Private Eye, 1968 (LYN 1677).

The Pop Way to Save, Post Office Savings Bank, c. 1968 (SFDK-18H).

The Beatles Sixth Christmas Record, 1968 (LYN 1743/1744).
Two-sided official Fan Club issued Christmas disc. This was the sixth in the series of seven important festive flexi discs issued by The Beatles, available only via the Beatles Fan Club. The original idea for Christmas flexi was to appease fans whose letters to The Beatles had not always been answered in good time. This particular issue from 1968 features improvised music, talking, poetry and impressions by The Fab Four. Also included is a very fine performance of 'Nowhere Man' by Tiny Tim, which was recorded by George Harrison in New York. Apparently The Beatles did not want to get together and record a special Christmas message in 1968 – they'd just finished the marathon that was to become *The Beatles* (*The White Album*), so various recorded snippets were supplied to friend and broadcaster Kenny Everett, who brilliantly edited these into a two-sided whole in his classic cut-up concrete style. The psychedelic artwork was uncredited, but features a girl with Kaleidoscope eyes. All of The Beatles flexis were reissued by Apple Records in 2017 as a box set entitled *The Beatles Christmas Records*.

CHRISTMAS
1968

The Bread Band, Hovis, 1969 (SFDK 13H).
Trade flexi by Hovis in die-cut gatefold. Instrumental extended jingle – wheat with a trad beat. Pictured: front and back of sleeve.

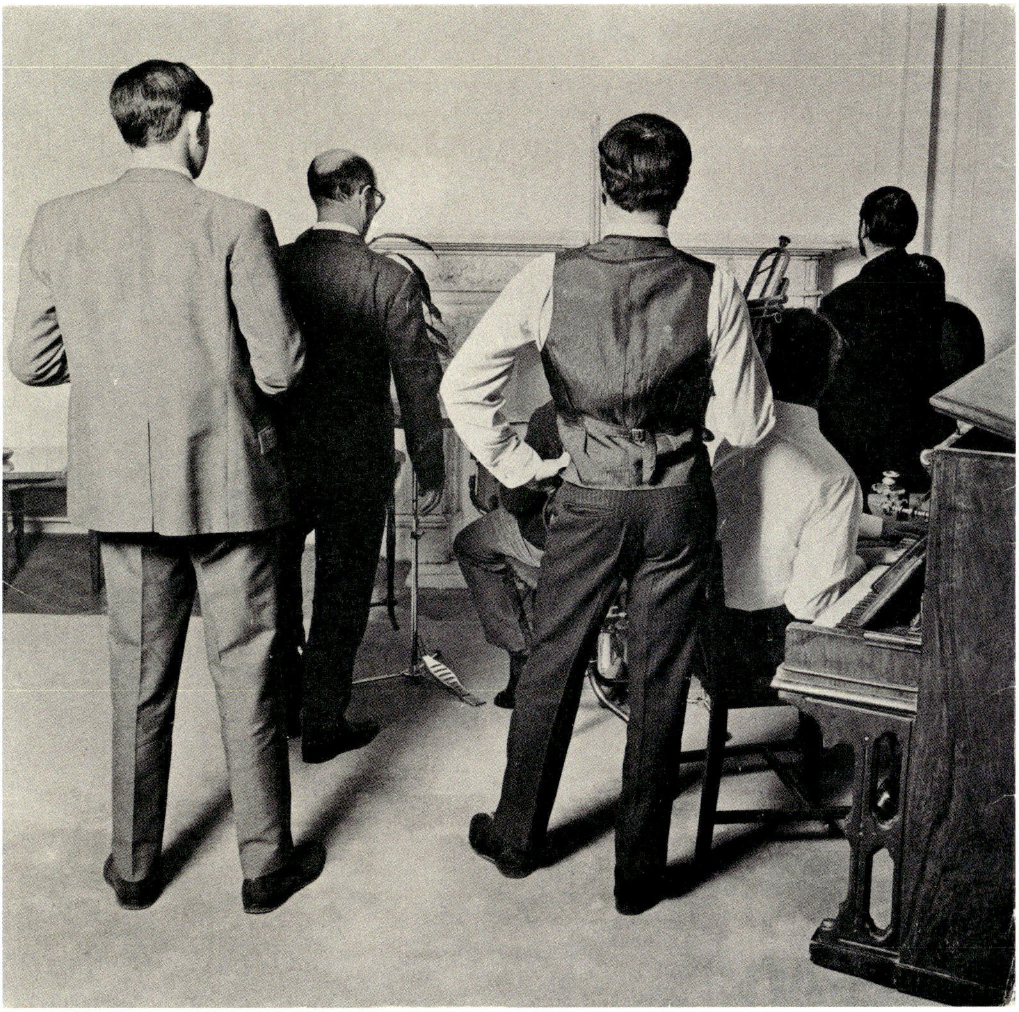

Michael Aspel chats up the Barclaygirls, Barclays, 1971 (LYN 1979).
One-sided recruitment disc. The smooth talker interviews ten girls about their new careers at Barclays. Includes a job application form.

Ready Brek, Lyons, 1971 (SFI 83).
One-sided, with Noel Edmonds introducing two cues from Warner Bros: i) Sasha Distel from his new album *Sasha* and ii) Johnny Harris from his groundbreaking *Movements* LP.

Grand Prix, Findus Frozen Foods, 1972, (LYN 2486).

100mph
Copse
A – Starter Button.
B – Oil Pressure. Oil Temp.
C – Rev. Counter.
D – Fuel pressure. Water Temp.
E – Fire extinguisher switch.
F – Ignition master switch.
G – Ignition switch.
H – Fuel pump switch.
I – Fire Extinguisher.
J – Kill switch (Turns off ignition in an emergency)

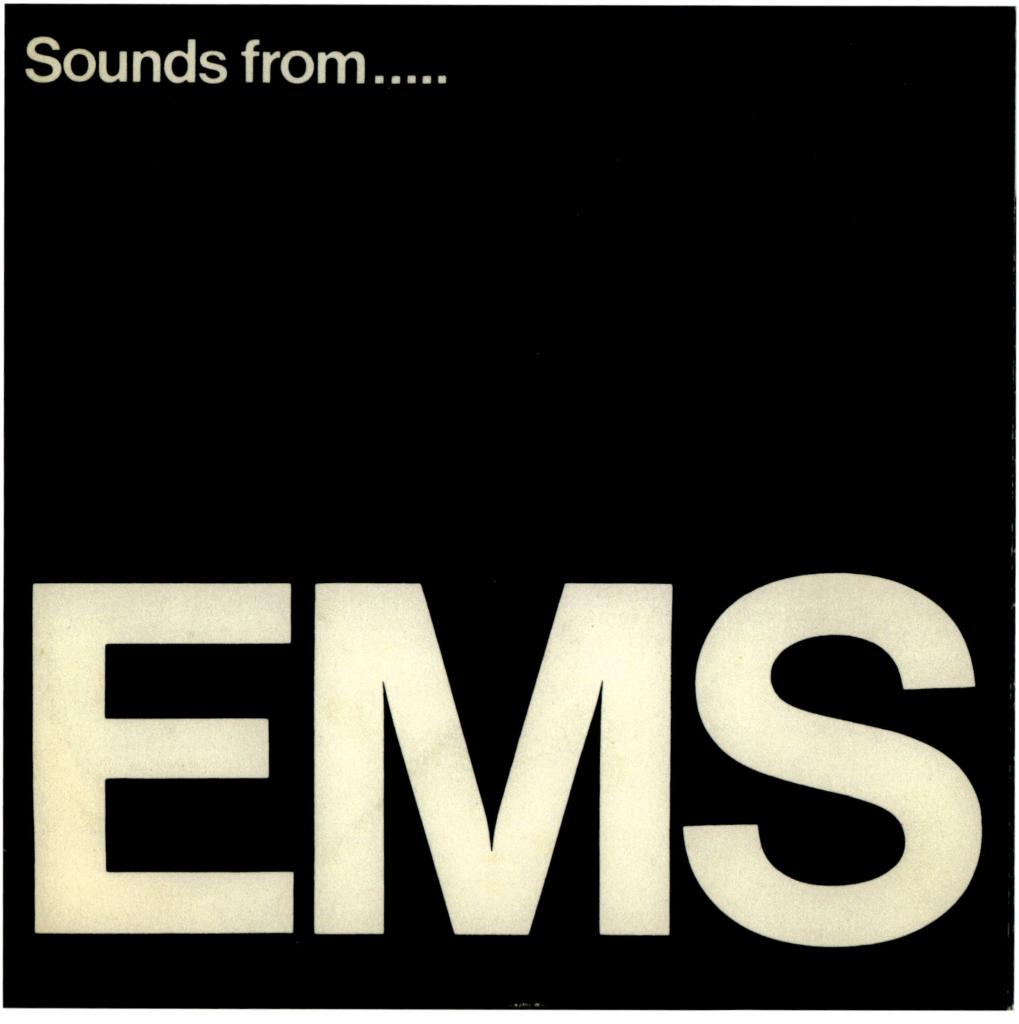

Synthi, Sounds from... EMS, Electronic Music Studios, 1972.
Legendary two-sided demo disc from EMS Synths, circa 1972. Includes music by leaders in the British electronic music field: David Vorhaus, Dudley Simpson, Delia Derbyshire etc, introduced and explained by Tristram Cary, one of the EMS founders.

SYNTHI SYNTHI SYNTHI SYNTHI

Brain Salad Surgery, Emerson Lake & Palmer, NME, 1972 (LYN 2762).
Perfect miniaturised replica of the classic prog rock album. This was originally given away free with the 3 November 1973 issue of the *New Musical Express* – two weeks before the actual release of the full album. This unusual and great little die-cut object must have acted as a most exciting advanced promo for what was to be ELP's fourth album – which charted at Number 2 shortly after release and remained in the charts for a further 18 weeks. Once opened up, the iconic H. R. Giger sleeve reveals a one-sided flexi cut at 33 rpm, which plays the title track, followed by short segued excerpts of all the other tracks on the album. According to rumour, the expression 'Brain Salad Surgery' is possibly a euphemism for fellatio.

'Fight The Flab' with Terry Wogan, Playtex, 1972 (SFI 117).
One-sided light-hearted disc from Playtex bras and girdles, encouraging exercise to shape-up.
'Fight The Flab' was a regular spot on Wogan's Radio One show from 1970.

Lipsmackinthirstquenchinacetastinmotivatingoodbuzzincooltalkinhighwalkinfastlivinevergivincoolfizzin, featuring Kenny Everett, Pepsi, 1973 (LYN 2464).

One-sided trade promo for the groundbreaking and hugely successful 1973 Pepsi Cola advertising campaign. The disc, introduced by Kenny Everett includes some average comedy, excerpts from the TV commercials plus some scant information about available point of sale. The origins of this classic campaign go back to the young creative Dave Trott. UK-born Trott has studied at The Pratt Institute and had arrived back in London as a junior copywriter working at Boase Massimi Pollit. When the Pepsi brief arrived on his desk it included a long series of words and phrases that Pepsi wanted used to promote their drink in the new TV advert, such as 'cool', 'lip-smacking' 'thirst-quenching' etc. Trott had just read Tom Wolfe's first book *The Kandy-Koloured Tangerine-Flake Streamline Baby* and quickly realised that he could simply put all the words Pepsi wanted in the commercial together in one long line. He did, and by doing so wrote himself into advertising history.

Lipsmackinthirst
quenchinacetast
inmotivatingood
buzzincooltalkin
highwalkinfastliv
inevergivin
coolfizzin

PEPSI

Sweet Taste of Life, Hermesetas, date unknown.

Shake A Tail Suzy, Barry Sheene for Suzuki, 1973 (SFI 144).

Hullo Sailor, Private Eye, 1972 (LYN 2558).
One sided x-rated comedy/satire by Private Eye – 1972

Teach Yourself Heath, Monty Python, 1972 (SHOL 1547).
Included with the original release of *Monty Python's Previous Record*. Introduced by Michael Palin, the actual lessons in Heath are all given by Eric Idle. It cleverly uses recordings and cut-ups of speeches by Edward Heath, focusing in on his distinctive ways of saying words like 'our', 'now' and 'history', repeating them and getting you to practice along. The track was included in a 2014 box set: *Monty Python's Total Rubbish*.

Easter Greetings to You, Purley Christian Outreach, c. 1973 (SFI 146).

On Love, Kenny Everett & Michael Aspel, White Elephant Publications Ltd, c. 1974 (SFI 210).
Begins as a spoof of a porn flexi, settles to Everett-style broadcast and educated chat on love.

Great Party Hits, Barker & Dobson / Bensons, 1973 (Sonopresse 2100).

Drinking Song, Double Diamond, 1973 (Sonopresse).

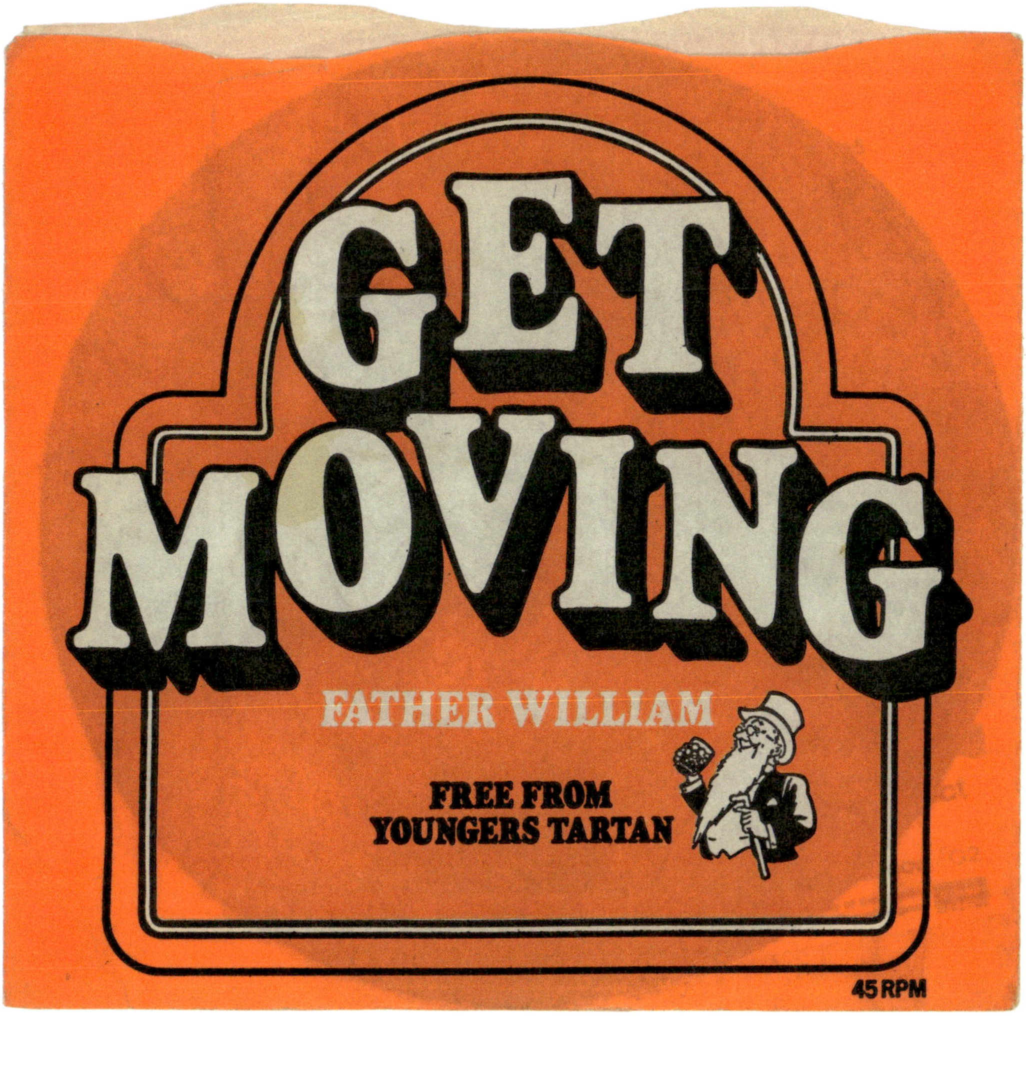

Get Moving, Father William for Youngers Tartan, 1974 (LYN 2596).

Mystery Voice, Lever Brothers, 1973 (SFI).
Lever brothers promotional disc pressed on odd 'Polyart' plastic.
The mystery voice is almost impossible to hear as the pressing is so very poor.

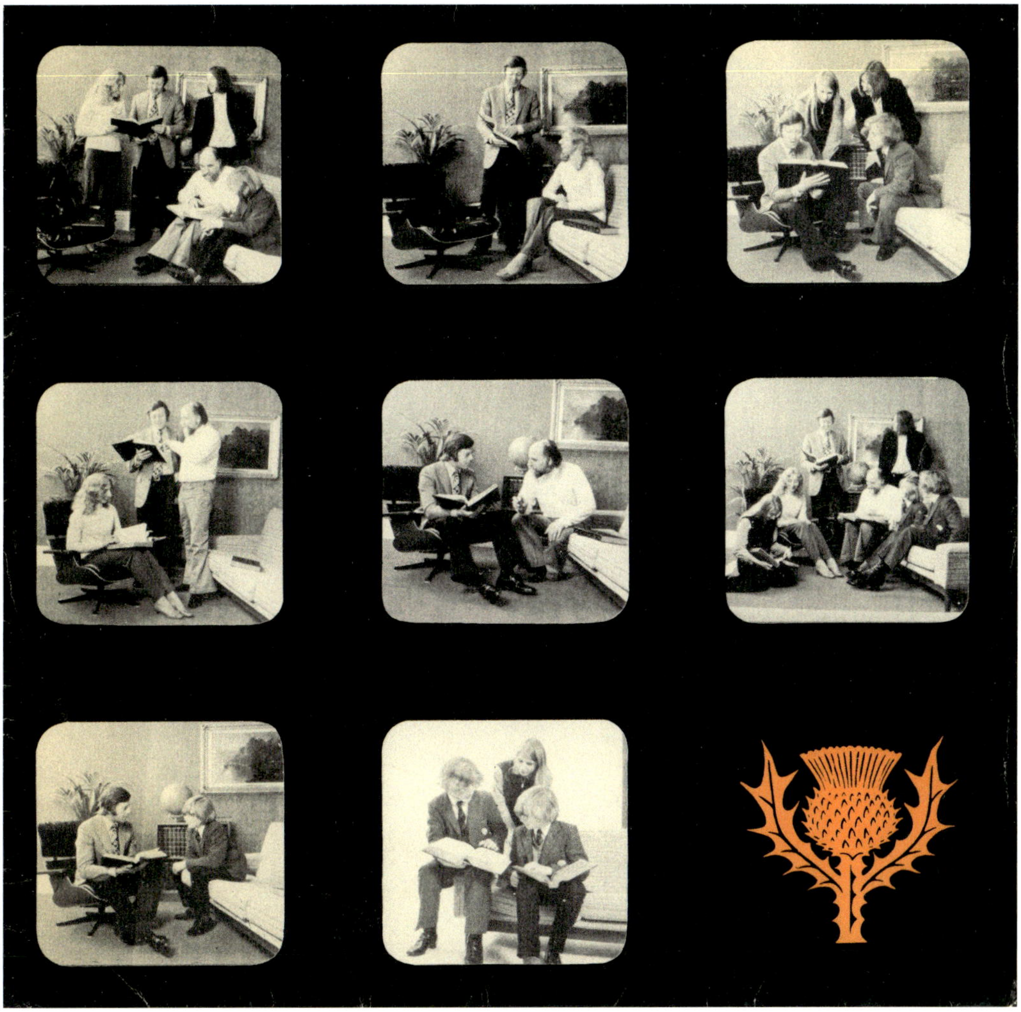

Encyclopaedia Britannica, Michael Aspel, 1970s (PC 053/054).

The Morecambe and Wise Happy Birthday Record, Maltesers Treets Revels, 1974 (LYN 2809).

Listen to Your Career, Magnus Magnusson Interviews Peter Snow, Director of Retailing, J. Sainsbury Limited, 1975 (LYN 3048/3049).

Sounds of the Sea, Ribena, 1976 (LYN 3248/3249).
Michael Aspel introduces a two-sided sea trip with Ribena.
Features a rather annoying shanty, some educational interjections about birds and wildlife and SFX providing some well-needed atmosphere.

Garden Bird Songs, recordings by John Kirby, RSPB, 1976 (LYN 3476/3477).

Saturn and Jupiter, keyboard demonstrations played by Brian Sharp, 1976, (LYN 3593)
Oddly unbranded two-sided demonstration disc highlighting features of two electronic keyboards played by Brian Sharp. Easy listening, but not very inspiring.

JUPITER 2
DEMONSTRATION

I'm The Noodle Doodle Man, Jon Pertwee for Heinz, 1976 (LYN 3611/3612).
Two-sided, with John Pertwee singing as the Noodle Doodle Man, covered in Heinz Tomato Sauce.

A DAF *Trucker's Romance*, 1978 (LYN 5624).
One-sided. Instantly forgettable country and western number about DAF trucks.

Take Advantage / Get Yourself Together, Bob Rogers & Playground, Express Dairies, 1977 (SFI 69).

Noel Edmonds Finds Out About Barclays, 1977 (SFI 287).

Fulham Fallout Firty Free, The Lurkers, free with *Ain't Got A Clue*, 1978 (BEG 6 ½).

Top Man, 1978 (LYN 6595).
Wolf-whistling one-sided promo about 'bloke ogling' – the magic effect apparently brought on by wearing Top Man clothes.

Central Milton Keynes, 1979 (SFI 463).
One-sided song, written about Central Milton Keynes. Apparently 'You've Never Seen Anything Like It'. Sublime in many, many ways.

Springtime for Hitler, Legs 'Larry' Smith/*The Inquisition Song* (excerpt), Mel Brooks, for *Event* magazine, 1981 (LYN 10286).

Leather Sex Orgy, 1981.
One-sided freebie disc from *Rapier* magazine. Unconvincing grubby chat between two ladies, leading to even less convincing dominatrix scene with a male 'slave' client.

The Royal Romance, 1981 (Flexi Records).

The Months Ahead, Dr. Miriam Stoppard, Health Education Council, 1980s (Sound for Industry).

Coronary Prevention at Work, Trafford Health Authority (Flexi Records).

We Can Help, David Hamilton for Provident, 1982 (LYN 10972).

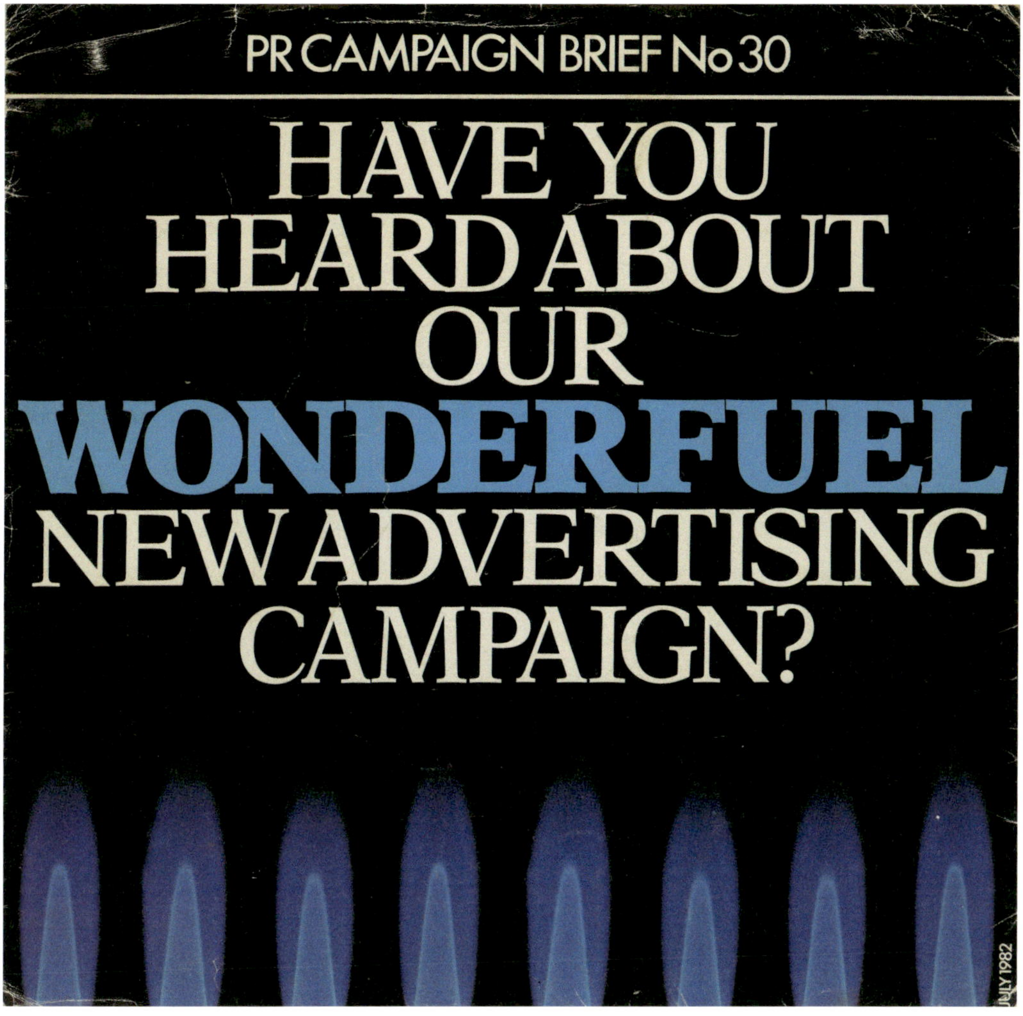

Wonderfuel Gas!, Madeline Bell for British Gas, 1982 (Flexi Records).

Jonny Ball Introduces Science, 1981 (Salter Science 763).
Free from one of five junior science sets by Salters. An engaging light hearted one-sided science talk by the charismatic Mr Ball. Some odd electronic dabblings in there too.

Elimination Dancing / King Kong, Bow Wow Wow, 1982 (LYN 11358).

ZX-81, VIC-20, *Spectrum Games*, *Your Computer*, 1982 (Flexisoft).
Issued with *Your Computer* magazine in 1982, the B side of this 'Flexisoft Disc' plays the recordable basic code for the game *Othello* for the ZX-81.

The Lady Lies, Genesis, Flexipop 021, 1982 (LYN 11666).

Message from Robert Maxwell, British Printing & Communication Corporation, c. 1982.
One-sided personal message from Robert Maxwell.
On white plastic, with possibly some little white lies along the way.

You and Your Bendix, Valerie Singleton for Thorn Domestic Appliances, c. 1982 (LYN 11859).

Fermaley Flame Sings For Your Birthday, 1982 (LYN 12213).
Manic birthday song, to John. Rhymes 'birthday' with 'earthday', which is cool.
Not sure if Fermaley Flame is a guy, a girl or just a flame. One-sided.

The Bride Wore Blue, Barbed Choir, This End Up Records, 1983 (LYN 12776).

Have You Heard The Noise, Golden Wonder, c. 1982 (SFI 709-352).
One-sided disc promoting new and old flavours from Golden Wonder,
plus the classic soundtrack to the 'Britains Noisiest Crisp' campaign.
Annoying lack of punctuation on the front sleeve.

Match Mastermind of Soccer, early 1980s (FLX 177, 178, 179).

MEB *Switched on for You*, Midlands Electricity Board, 1983 (LYN 14609).
One-sided appropriate camp electro pop number,
repeating the simple phrase 'Switch It On Switch It On Baby'.

Jobs For A Change, Ken Livingston and the Flying Pickets, 1984 (LYN 14479).
One-sided. 'Ken Livingston' promotes the GLC employment push.
With World Music vibe. Slight 'rap' feel.

Xmas Cuts, Island, 1984 (A Reccard).

I Feel Free, Jack Bruce / Peter Brown, Renault, 1986.

Stay In Bed/Four Letter Girl, The Shamen, for Skipping Kitten, 1986 (WRY 002/SF 846).

The Oink Song, The Oinkletts/*Oink Rap*, Uncle Pigg, Oink, 1986 (In Tape IT 032). One-sided manic post punk/electro hip hop style numbers composed by Mark Riley and chums. Free with first issue of *Oink* magazine in 1986. Well, no one would have paid for it.

Invitation, Fisher A/V, 1987.
One-sided disc with odd and terrible big band electro sound, backing a shouty man talking about Fisher's audio visual showcase in 1987.

Skeye Flexi-Dish, Private Eye, 1989.

Raft, Screeming Custard/Drive Away, Dead Famous People,
Bring On Bull Records, 1992 (BULL 1-0).

Audacious 2, 1993.

Four Corners Irregulars
A series of books presenting a visual history of modern British culture.
This is book 6.

Titles include:

1. Eyeball Cards, the Art of British CB Radio Culture
2. UFO Drawings From The National Archives
3. Poster Workshop 1968–1971
4. Leeds Postcards
5. Face In The Crowd

...with further volumes in preparation.

Set in Starling and printed on Garda Matt Art.

Published in 2019 by Four Corners Books
56 Artillery Lane, London E1 7LS

Designed by John Morgan Studio
morganstudio.co.uk

Print production by Martin Lee
Reprography by Martin Chapman
Printed in Italy by Printer Trento

Distributed in the UK by Art Data
artdata.co.uk

ISBN 978-1-909829-14-5

Slight 'rap' feel at
fourcornersbooks.co.uk

Back cover and endpapers photography by Mikael Gregorsky.

Many thanks to the following collectors who helped us put the book together: DJ Food (djfood.org), Andrew Beedell Coram, Shane Quentin (gardenofearthlydelights.com), Jez Randell (allvinylexperience.com) and Jonny Trunk (trunkrecords.com). Special thanks also to Sandra Lynton.

THE RECORDED WORK RESERVED UNAUTHORISED PUBLIC PERFORMANCE BROADCASTING AND COPYING OF THIS
FAST
PRODUCT